No Dead Fish for Ginger!

To Teddy,
Have fun reading
about dolphins!
Cathy Marine

The Story of a Sarasota Bay Dolphin

by Cathy Marine

Eifrig Publishing LLC

Lemont   Berlin

Published by Eifrig Publishing, LLC
PO Box 66, 701 Berry Street, Lemont, PA 16851, USA
Knobelsdorffstr. 44, 14059 Berlin, Germany.

For information regarding permission, write to:
Rights and Permissions Department,
Eifrig Publishing, LLC
PO Box 66, Lemont, PA 16851, USA.
permissions@eifrigpublishing.com, 888-340-6543

Library of Congress Control Number: 2011941476

Marine, Catherine
No Dead Fish for Ginger! The Story of a Sarasota Bay Dolphin
by Cathy Marine
p. cm.

Paperback ISBN 978-1-936172-44-3
Hardcover ISBN 978-1-936172-52-8

[1. Marine Biology, 2. Animals-Dolphins]
I. Marine, Cathy, author. II. Title

Summary: Ginger is a Sarasota Bay dolphin who stranded in December 2008. She was rescued,
rehabilitated and returned home by the staff and volunteers at Mote Marine Laboratory. For
40 years, her family has been part of the world's longest running study of a dolphin population.

*Proceeds from the sale of this book will benefit dolphin research and the Dolphin and Whale
Hospital at Mote.*

16 15 14 13 2012
5 4 3 2 1

Printed on acid-free paper in the USA ∞

TABLE OF CONTENTS

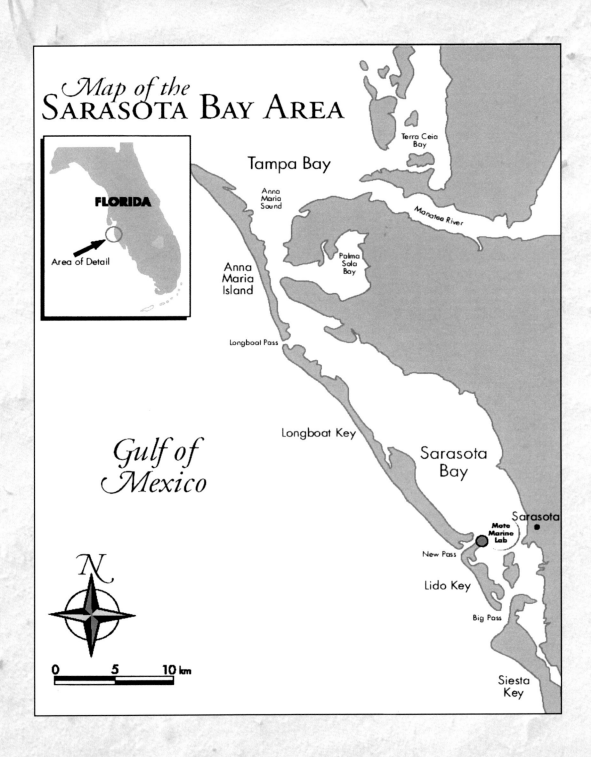

Map of the
SARASOTA BAY AREA

FLORIDA

Area of Detail

Tampa Bay

Terra Ceia Bay

Anna Maria Sound

Manatee River

Anna Maria Island

Palma Sola Bay

Longboat Pass

Gulf of Mexico

Longboat Key

Sarasota Bay

Sarasota

Mote Marine Lab

New Pass

Lido Key

Big Pass

N

0   5   10 km

Siesta Key

"There's a dolphin on the beach."

The news spread quickly among the early morning beach walkers on Siesta Key. Some people raced home to get towels, sheets, and plastic bags. Others headed over to the dolphin to see what they could do. One person called the special phone number posted on many Florida beaches to report this animal emergency.

The telephone call reached the Mote Marine Laboratory stranding team. They started the network of calls that would bring in the hospital staff, the Sarasota Dolphin Research Program and trained volunteers who would help with the dolphin's rescue.

5

Dr. Randall Wells, head of the Sarasota Dolphin Research Program (SDRP), lives just up the beach from where the animal was located, so he raced over as soon as he got the phone call. He was the first of the Mote Marine Lab staff to reach the stranded bottlenose dolphin.

Randy was pleased to see that the people who had discovered the dolphin had done everything exactly right. Sometimes people who find a stranded marine mammal try to push it back into the water. This can result in injuries to the people as well as to the animal. It takes a trained professional to determine whether a dolphin is healthy enough to be returned to the water or if it needs to go to a dolphin hospital.

About ten early-morning beach walkers had gathered around the dolphin. They had covered the bottlenose with a wet sheet, making sure that the blowhole was clear, and they were shading it with umbrellas. Dolphins can overheat quickly when they are out of the water, and they can sunburn. Even the early morning rays of the December sun can cause severe damage to their sensitive skin.

Using clean, clear plastic bags, people were keeping the dolphin wet with water from the Gulf of Mexico. This would keep her skin from drying out. They made sure no water went into the blowhole. (That would be like someone dripping water up your nose).

The dolphin was in a small trough filled with water, and someone had dug holes in the sand under the pectoral fins to relieve the pressure there. Gravity is hard on stranded marine mammals. In the water, pressure on their bodies changes depending on the depth at which they are swimming. Water pressure increases as you go deeper. That's why, if you dive to the bottom of the pool, sometimes your ears pop. No matter how deep or how shallow the dolphin swims, however, the pressure is the same all around its body.

When a dolphin strands, the pressure on land is very different. The belly and the fins are squashed against the sand, and the weight of the animal presses down on its internal organs, resulting in a lot of damage if the pressure is not relieved quickly.

When Dr. Wells arrived on the scene, he helped dig the trough deeper and increased the water level to reduce the pressure on the dolphin's body.

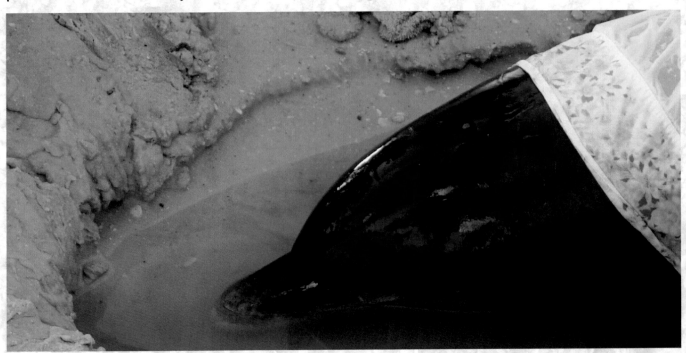

Another Mote staff member, Connie, who also lives nearby, arrived soon after Randy. She began to monitor the dolphin's breathing and heart rate while Randy called additional staff and then began answering questions from the curious crowd.

"Why did she strand?"

There are many reasons why a marine mammal may strand including sickness, injuries, being chased ashore by a predator, or, in the case of a group of dolphins, following a sick individual onto the beach. In all cases, something is wrong and the situation needs to be evaluated by a person who knows about dolphin biology and behavior.

Later, some would suspect that this three-year-old dolphin had stranded because of inexperience. She might have been chasing fish through the shallow water just off the beach, perhaps gathering a yummy breakfast, and she misjudged the extremely low tide. The beach area where she often caught her prey was shallower than usual, and, before she knew it, she was too far onto the sand to turn around.

"What kind of dolphin is this?"

Randy talked about dolphins in the Gulf of Mexico. There are over 30 species of dolphins. Most of the near-shore (as opposed to deep-water) Gulf dolphins are common bottlenose dolphins. The Latin name is *Tursiops truncatus*. They are called "bottlenose" because of the short cylindrical beak or rostrum, and they are found in tropical, sub-tropical, and temperate waters throughout the world. They are common in coastal areas.

In addition to the shape of the rostrum, a bottlenose dolphin is usually dark to medium gray on top in the dorsal area, and lighter as you go down toward the belly.

Randy pointed to the color difference on the stranded animal. Her dorsal fin is tall and recurved, or "falcate" with the tip pointing toward the tail flukes, and the pectoral fins on either side of the dolphin's body are longer and wider than many other dolphins.

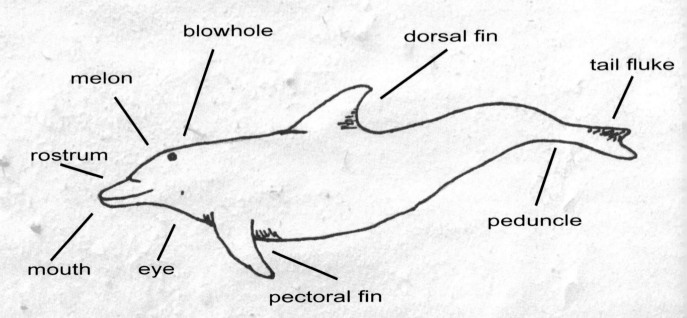

10

"What is the difference between a dolphin and a porpoise?"

Dolphin teeth are cone-shaped. Porpoises have spade-shaped teeth and their dorsal fins are more triangular. If you could count this animal's teeth you would find 20 – 25 on both the upper and lower jaws on each side. And dolphins can bite so it is important not to get too close to those teeth.

A wild dolphin from Sarasota called "Beggar" has sent people who tried to feed him to the hospital with fairly serious wounds. Feeding wild dolphins is illegal and bad for the animals.

By now, the Mote stranding team had arrived with a truck that serves as an animal ambulance. The back of the truck has stretchers and poles to carry a sick or injured marine mammal, thick foam pads to support it, and buckets and sponges to keep the animal wet. Lynne, a member of Mote's veterinary staff, brought the "crash kit," a case with medical equipment and supplies for use if the dolphin stopped breathing or went into shock.

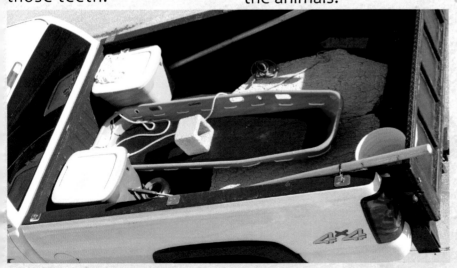

11

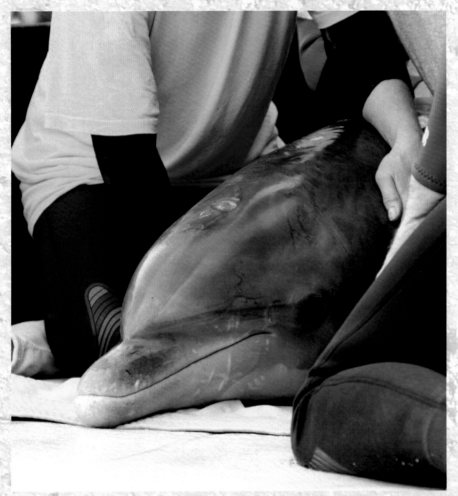

initial diagnosis was that she needed medical intervention.

When the dolphin was ready to be transported to Mote's Dolphin and Whale Hospital, Mote staff raised her slightly to move the stretcher under her, then lifted her carefully and carried her to the truck.

Once the animal was placed on the foam pads, the staff was positioned along both sides of the dolphin for support. Buckets were filled with water so staff could sponge water over her to keep her wet, while the truck drove nine miles north to the hospital.

Lynne checked the dolphin carefully. As the trough underneath her was deepened and filled with more water and her body was adjusted so her dorsal fin was upright, the animal's respiration rate decreased from 3 to 5 times a minute to 1 to 2 times a minute, closer to normal. She appeared to be calm and had only a few minor cuts on her tail, but her lungs sounded "rattley." The

WELCOME TO THE HOSPITAL

Back at Mote, volunteers and staff were getting the Dolphin and Whale Hospital ready for its new patient. There are a dozen small pools, primarily for turtles, and two large critical care pools, 30-ft in diameter and 9-ft deep. One of the large pools would be the new home for the incoming dolphin. A loggerhead turtle had recently moved out of the pool, so the turtle toys had to be removed and the tank was filled with water from a nearby waterway. By using natural seawater, the temperature and chemical composition would be the same as what the dolphin is used to.

Once the pumps to fill the pool were started, team members got out foam pads and canvas stretchers to support and transport the dolphin, and buckets and sponges to keep it wet until it was lowered into the pool.

The pools are surrounded by a deck and a fence. A crane would lift the dolphin from the truck, over the fence onto the deck and, finally, into the tank. A scale would be attached to the crane to weigh the dolphin when the animal arrived.

An assortment of medical equipment had to be organized so veterinary staff could take blood, fecal (poop), and urine samples. These would show infection or parasites in the digestive system. In addition, the urine can indicate kidney problems. A swab of the blowhole would enable vets to check for infection or parasites. Because most of the patients are dehydrated when they arrive, plastic containers of Pedialyte® and plastic tubes to get the liquid into the dolphin's stomach, a procedure called "tubing" the dolphin, were ready in the vet lab. Since dolphins do not have a gag reflex, gently sliding a narrow soft rubber tube through the mouth and the esophagus into the stomach can be done with minimal discomfort to the animal.

A new notebook was prepared so staff and volunteers could start recording everything that was done by and to the dolphin from the moment it reached the hospital. *What noises does the dolphin make? How often does it breathe? Is it eating? If so, what and how much?* Keeping records is extremely important.

Veterinary staff relies on these records as they decide on diet, medications, and animal care. The National Marine Fisheries Service (NMFS), the federal agency responsible for monitoring the rehabilitation of marine mammals and determining the future of rehabilitated dolphins, will require information from the notebooks.

"Truck is coming across the bridge. Get your wetsuits on."

Those who were ready started downstairs to meet the truck. Some of the volunteers have been doing this for many years. Others have only had a few experiences with marine mammals. All volunteers have been through several

crane lifted her to the deck, volunteers and staff raced upstairs. Lynne quickly measured the animal from the tip of the rostrum to the tail, and her girth behind the blowhole and behind the pectoral fins. Then the dolphin was lifted again, and volunteers and staff started down the ladder into the tank.

Throughout the transfer, one volunteer called out "blow" every time the dolphin breathed and another recorded the "resp," short for respiration, in the notebook. It was quiet. This is a hospital so voices are low and talk is kept to a minimum.

days of training, and understand that this is a wild animal. Dolphins are cute. They always look like they are smiling. But a dolphin can be dangerous – especially a sick dolphin. A flick of the muscular tail can break a person's ribs. This animal is over 250 lbs (123 kg) and can do some real damage if everyone does not work together on the transfer from truck to tank.

All went well. The dolphin was moved to the sling; then, as the

F127 - Ginger's mom

By now, the Sarasota Dolphin Research Program team had had a chance to compare photos of this dolphin to their database of over 300,000 photos of Sarasota Bay dolphins. The nicks and notches on the dorsal fin, perhaps from contact with other dolphins, were unique and Aaron verified that this was 1271 – "127" is her mother's identification number, and "1" because she is F127's first calf. In order to process her medical information and make communication easier, staff named her "Ginger," short for gingerbread, a December holiday treat.

After measuring her weight, Ginger was lowered into the pool that would be her home for the next eight weeks. For the first two hours people held Ginger and walked her around the tank. She needed to understand that there are boundaries to her new home. If she swam too fast, she could run into the side and injure herself. The staff also needed to make sure she was able to swim. Often animals arrive at the hospital so sick that they must be supported in the water for several days before they are ready to swim without help.

By lunchtime, Ginger was ready to try swimming on her own. Volunteers stood with their backs to the wall of the pool and circled the area. When staff released Ginger, each person moved to make sure the dolphin did not smash into a wall. Within minutes it was clear that this animal had figured out her new surroundings.

Volunteers quietly climbed out of the tank. Some took off their wetsuits, showered and left the hospital. Others remained ready to return to the pool if Ginger's condition were to change. Those leaving stopped on their way out to check on Ginger one more time. It had been an exhausting morning.

HEALTH EVALUATION

By mid-afternoon, the results of the blood work were in from Sarasota Memorial Hospital. The hospital had done an assortment of tests on small samples of Ginger's blood, just as is done for humans when they are admitted to the hospital. As expected, Ginger was dehydrated and seemed to have respiratory problems, probably pneumonia, a common illness of stranded dolphins. Vet staff continued to give her Pedialyte®, and started her on antibiotics. They conducted a full exam and gave her a shot of vitamins.

After the exam, Ginger was offered her first meal. Because this animal would

Taking a blood sample from the tail fluke

be returned to the wild once she recovered, every effort would be made during her stay at the Dolphin and Whale Hospital to keep her from associating humans with food. No one wanted her to beg from fishermen or boaters. This is a big problem throughout the southeastern United States, leading to dolphin deaths and injuries, and angry fishermen. With that in mind, dead capelin and herring were thrown over the side of the pool. Volunteers watched from a distance to see if she ate the fish.

19

No. Not interested. She refused to eat the dead fish. They tried throwing in live fish. She still was not interested. Volunteers used nets on long poles to scoop the fish out of the tank.

The next morning, the vet staff did another exam and tube fed Ginger, giving her gruel – a "fish smoothie" of herring and Pedialyte®. Again, they tried throwing fish over the side of the tank, but Ginger refused to eat any of the fish that were thrown in. Ginger was tube-fed three times a day for the first six days, but after one day of fish smoothies, she was having problems with gas so staff stopped adding the herring gruel to the Pedialyte®.

The third day, when Ginger was offered dead fish and live fish, primarily pinfish and mullet, she ate 15 of the live fish. For the next three days Ginger continued to get a mixture of dead and live fish. She ate only the live fish – and not all of those. Staff tried to trick her into eating the dead fish by throwing in a dead fish right after she had eaten a live one. Ginger did not fall for the trick. She would only eat the live fish .

In the wild, most dolphins eat only live fish. Maybe they have learned that dead fish may be toxic or may contain chemicals or parasites. Ginger was born during a period when Sarasota Bay experienced a major "red tide," an algae bloom that caused sickness and death of many fish, sea turtles, manatees and dolphins. Maybe Ginger's mom taught her to avoid eating dead fish to protect her from this deadly toxin.

On the sixth day, Ginger ate 45 live finger mullet. Staff decided that because Ginger would probably get well and be returned to Sarasota Bay soon, her hospital diet should be like her meals in the wild.

While mullet are part of the Sarasota Bay dolphin diet, the bottlenose dolphins often choose pinfish. Since these fish are easier to obtain (they are often sold at bait shops) and cheaper than the mullet, pinfish become the main entrée in Ginger's diet.

Ginger was a picky eater. If the fish was too small, under 0.06 kg, she refused to eat it. Maybe it was just not worth the energy it takes to catch such a small fish. If the fish did not swim fast enough or looked weak, she refused to eat it. She liked finger mullet and silver mullet but refused to eat black mullet. She would eat ladyfish, blue runners and sea bass, but not whiting. She liked to play with her food and sometimes chased a fish until it died – and she wouldn't eat a dead fish! After she was fed, Ginger often did a lot of tail slapping. Did this mean she wanted more?

That she would prefer a different menu? Whatever the reason, it definitely meant she was not happy!

The dolphin continued to lose weight. Staff adjusted food amounts and feeding schedules to try to fatten her up and get her healthy again. Some days she ate a lot; some days she just ignored the food. Staff continued to check stomach contents and blood work to see if there was any medical reason for her to refuse the food.

In the meantime, Ginger was back to tube-feeding. After twelve days in the hospital, staff was trying to offer Ginger food every two hours from 8:00 am until midnight in an effort to get her to gain weight. They continued to adjust her diet until, on January 9, after 24 days in the hospital, she finally began to gain weight. A week later, Lynne reported that she was "cautiously optimistic." Ginger was gaining weight, her blood work had improved, and her lungs sounded much better.

After another week, Ginger was taken off all medication. She was now eating up to 35 fish every three hours from 8:00 am until midnight. That was over 200 fish a day!

The weather had gotten colder, so the larger fish disappeared from the Bay. That meant more pinfish – the fish the volunteers and staff had come to hate because of the sharp spines on their dorsal fins.

FOOD, GLORIOUS FOOD!

For most patients at the Dolphin and Whale Hospital, the fish truck brings in frozen fish twice a month. National Marine Fisheries Service (NMFS) regulations require that the fish be "restaurant quality." So, the truck that delivers fish to many of the restaurants in Sarasota also makes a stop at Mote.

Ginger, however, would only eat live fish. This presented a new challenge to the veterinary staff. Where can you get 200 live fish every day, seven days a week? The answer was a bait shop about two miles from the hospital.

The feeding was very time consuming. The veterinary staff needed to know exactly how much Ginger was eating, so each fish had to be weighed and measured before it was tossed into the tank. That meant netting the fish from the bucket, putting it on the scale, writing the weight on a chart, picking up the fish again, and running over to the tank to throw it in. In order to prevent Ginger from

associating humans with food, the thrower had to stay out of sight -- so the thrower's aim had to be pretty good.

If Ginger did not eat the fish, the fish weight had to be taken off the chart, so volunteers peeked under the tank's screen and signaled to the recorder when Ginger had eaten the fish. If she did not eat the fish, another was weighed, recorded and thrown in

23

the tank. If she rejected three fish, feeding was stopped. If the fish died or if after ten minutes Ginger had not eaten the fish, they were removed from the tank. When each feed got up to 35 fish, it often took over an hour to feed Ginger.

After several days of pinfish, the vet staff noticed sores in Ginger's mouth. Dolphins do not chew their food. They have no jaw muscles for chewing. They use their teeth to capture and hold onto their prey and swallow it whole, or whack it against the water or rub it on the bottom to break it into chunks that can be swallowed. In open water, dolphins can catch the pinfish and flip them so that pins are pointed backward and do not injure the mouth.

Apparently, in the tank Ginger was not able to reorient the pinfish so her mouth was getting cut by the pins. The best remedy seemed to be to cut off the pins before throwing her the fish. This added one more step to the feeding process – and one more opportunity to be pricked by the spines.

During her 8-week stay at Mote, Ginger ate about 4000 pinfish at $1.00 per fish. She was a very expensive patient.

Pinfish, anyone?

It is easy to see how the pinfish got its name. The spines of these small fish are sharp, like pins. Picking up a pinfish is like picking up a pincushion with the pins pointing the wrong direction! Volunteers tried wearing plastic gloves, leather gloves, performing acrobatics with the net to keep from touching the fish. Nothing seemed to work.

By the time the pinfish had been purchased at the bait shop, moved to a tank at Mote, moved from the tank to the bucket, netted from the bucket, weighed and measured, then thrown in the tank for Ginger, the fish had been touched or netted at least six times. The fact that most of the pinfish lived until Ginger ate them is testimony to the resilience of these rugged little fish. From the volunteer standpoint, if you multiply 200 pinfish per day by six touches, you get a lot of pinpricks. No one needed a reminder to wash their hands, and the Band-Aid box was in constant need of refills.

HEARING

The ocean is a noisy place. Waves crash against the shore. Boats zoom through the water. Fish munch on coral.

Maybe because sound travels more than four times faster through salt water than it does through air, hearing is very important to marine life. Dolphins use sound to hunt, to navigate, and to communicate with other dolphins. Although dolphins do not have outer ears like we do, the structure of their middle and inner ears is similar to those of terrestrial mammals.

In addition to the jaw claps, chuffs, and squeals that humans can hear,

dolphins, like bats, hear and produce sound in the high frequency range, typically above 40,000 Hertz (cycles per second), called echolocation. Using echolocation, also called "sonar," the dolphin produces clicking sounds from deep within its head. The melon, a fatty organ that looks like the dolphin's forehead, acts like a lens and focuses the sounds. The clicks bounce off the object and send back an echo, which is picked up by the lower jawbone and passed through the middle and inner ears to the brain. The brain then processes the information revealing the location, size, shape, speed and density of the object.

While humans hear sounds from about 20 Hertz to 20 kilohertz (and only small children with excellent hearing can hear 20 kHz), a bottlenose dolphin can hear from about 75 Hertz to over 150 kilohertz. Humans and most other land mammals can hear sounds underwater, but we don't know where they are coming from. Dolphins are able to identify the direction from which a sound comes.

Scientists are very interested in the impact of oil drilling, pollution, and military sonar testing on the hearing of marine mammals. They also wonder whether hearing or hearing loss may have an impact on dolphin strandings. Because hearing is critical to the survival of Mote's dolphin patients, each animal has a hearing test shortly after it is admitted to the hospital.

Dr. David Mann, from the University of South Florida and an Adjunct Scientist at Mote, conducts the tests called "AEP" for Auditory Evoked Potential. Dr. Mann places a speaker embedded in a suction cup on the dolphin's lower jaw where the animal normally receives sound; then he puts a sensor with a transducer on the dolphin's head immediately behind the blowhole. A reference sensor, also in a suction cup, is placed several inches behind the recording electrode and slightly off center. Short tones of various frequencies and sound levels are played via a computer that then analyzes the electrical activity in the brain to gauge the animal's hearing abilities. Ginger passed her hearing test with flying colors. Her hearing is excellent!

FAMILY AFFAIRS

Ginger was born in the spring of 2005. She was her mother's first calf. Dolphin moms are pregnant for about twelve months. The calf is born tail first; then Mom quickly twists away to break the umbilical cord. The baby typically swims to the surface on its own for its first breath with Mom nearby.

For the first few days, sometimes up to a week, "fetal folds," from its birth, give the calf a wrinkly look. While swimming and diving are natural behaviors for

dolphins, it takes the calf a few days to learn to coordinate the activity. It is fun to watch as a newborn bobs up and down and, finally, plunges its head down into the water to follow its mother.

During these first days, scientists believe mother dolphins whistle to their calves almost continuously, teaching them to recognize their mothers by sound. Within its first few months of life, the calf develops its own "signature whistle." Like a name, this whistle will allow mom and calf to identify each other, and call to each other when they are separated.

For the first four to six months, Ginger stayed close to her mom, nursing frequently, sometimes every fifteen minutes. In order to nurse, her mom, F127, made sure the calf was right by her side in position to drink the short bursts of high fat content milk

mom squirted into Ginger's mouth. She may have continued to nurse until she separated from her mother at almost three years old, but by the time she was six months old, she had probably started to eat fish. She may have practiced her skills by chasing a small fish and throwing it into the air and then chasing it again.

During the summer of 2007, Ginger was often spotted with her mother and her grandmother, FB13, in the shallow waters of Robert's Bay or in the City Island grass flats, safe areas for a young calf.

By June 2008, Ginger was no longer seen with her mother. It appeared the two had separated. Unlike humans, young dolphins between the ages of three and six usually leave their mothers, sometimes just before the birth of a new sibling.

The juvenile years are a difficult time for dolphins. On their own for the first time, they are excellent targets for local predators, mainly sharks, and their skills in escape, feeding and even avoidance of boats, may still need practice.

Sarasota Dolphin Research Program scientists have found that in Sarasota Bay, often the juveniles or sub-adults, ages three to ten or twelve, form their own groups of males and females. This gives them added protection and opportunities to socialize. They are very active, just like human teens, chasing each other, playing with small fish or sea grass, and just hanging out.

Male dolphins may form lifelong bonds and become known as "buddy pairs." According to Dr. Wells, if one male in a pair dies, it can be very hard on the survivor. If he is lucky, he may find another male, perhaps another survivor from his old juvenile group.

Once they reach maturity, the female dolphins will begin to associate more with other females. These could be other young mothers, or a dolphin may, at this point, spend some time with its own mother, sisters or older females.

For the first six months after Ginger's release, she was usually found within a mile of her mom and her new sibling. At times she was with another dolphin, but she was often alone. Then in August, Ginger was seen with a group of juveniles. It appeared that she had found friends to hang out with.

Ginger's grandmother, FB13, died Sunday, July 26, 2009. She was 50 years old. Her body was discovered by Mote's Sea Turtle Patrol, a group of volunteers who walk the beaches in Sarasota County every morning during turtle nesting season. The Sarasota Dolphin Research Program had known this dolphin since August 1975. A necropsy, the animal version of an autopsy, was conducted on FB13. There was nothing to indicate that the cause of death was anything other than old age.

29

SWIMMING IN CIRCLES

It was a cold January night, and volunteers were arriving for the midnight to 4:00 am shift. Another volunteer had called at 11:35 pm to report that he and a buddy had caught a lady-fish and would bring it up for Ginger's midnight feed.

The first five minutes of every half-hour, a volunteer records respirations, "resps" or breaths. The number of times a dolphin breathes can be an indication of her health as well as a clue as to whether she is sleeping, playing, happy, or upset. When Ginger is happy and relaxed, she breathes

about every 30 - 40 seconds. Just after she has eaten, she breathes more quickly – sometimes every 15 - 20 seconds. She can hold her breath for several minutes so volunteers are not surprised when she stays underwater for one or two minutes. Ginger often breathes irregularly. Is that because she is in a tank? Is she being playful? Or is that just her normal breathing pattern?

Between midnight and 8:00 am, Ginger was not fed. The volunteers sat quietly out of her vision range counting resps or just making sure she was OK. The dolphin swam slowly counterclockwise around the tank. She ignored the dolphin toys.

There was a "rub line," a large rope hanging across the tank that she could use to scratch off dead skin. In bottlenose dolphins, scientists have found that the outermost layer of skin may be totally replaced every two hours. That's a lot of dead skin! Ginger, however, was not interested in playing with the rub line.

Does Ginger sleep? Like humans, dolphins have two connected brain lobes. Unlike us, breathing is not a reflex action for dolphins. We do not have to think about breathing. We just do it. Dolphins, however, breathe air, but live in water. If a dolphin were to automatically take a breath underwater it could drown. So dolphins have to think about breathing. They have to be conscious. Therefore, a dolphin cannot sleep soundly like a human. It might suffocate or drown. Current theory says that dolphins probably sleep half a brain at a time. That is, one half of the brain is sleeping while the other half is awake. Ginger's swimming slows. Sometimes her breathing rate decreases, but a loud noise or a tap on the tank startles her awake, if only for a few seconds. Then half her brain is back in sleeping mode while the other half tells her when to breathe.

A new group of volunteers arrived for the 4 - 8 am shift. Other than counting resps, the first couple hours were very quiet. Volunteers watched Ginger enviously, wishing they could let half their brains sleep.

By 6:00 am, the horizon was getting lighter and the pelicans began to feed in the Sarasota Bay waters. The dog walkers and joggers began arriving in a nearby park. The best thing about the early morning shift is watching the sunrise over the bay.

There were clouds on the horizon. Ginger dislikes rain. The sun tarp can be removed when it is raining, but staff learned that Ginger prefers to be protected from the raindrops. She chuffs repeatedly. A chuff is just a loud exhale through the blowhole indicating that she is not happy. It sounds like a short blast of air. Ginger seems to chuff for a variety of reasons that her watchers do not understand, but rain is definitely one cause for chuffing.

By now, Ginger was wide awake. She tail slapped – once, twice, three times. Slapping her tail against the water makes a loud splash. It is another way Ginger communicates that she is unhappy. Maybe she thought it was time for breakfast.

READY TO GO HOME

By 6:30 am, staff and volunteers were arriving for the bi-weekly exam. The crane was hooked up to the scale to weigh Ginger. She had been off medications for two weeks now, so the staff members had begun discussions with NMFS about a possible release date.

At 7:30 am, volunteers started down the ladder into the tank. They stood in a line and moved toward Ginger, narrowing the space until she took a breath. Then David called "OK" and those closest to the dolphin grabbed her, making sure her blowhole was clear and upright. Unlike some patients,

Ginger was usually easy to catch. It was only after she was caught that she flexed her muscular tail, sometimes lifting three or four volunteers up off

the floor of the tank. She seemed to know how many people had hold of her, and the fewer there were, the more she struggled.

Once Ginger was stabilized, Lynne took blood samples from a blood vessel in the dolphin's tail fluke. Then she collected a fecal sample and a urine sample. Tubing Ginger would provide a gastric sample to check her stomach contents, and finally Lynne waited for a breath to get a swab of the blowhole.

This time, after weighing Ginger, the stretcher was hoisted over the wall of the tank and onto the deck. In anticipation of her release, Ginger would be freeze-branded. To freeze-brand a dolphin, 2-inch metal numbers are super-cooled in liquid nitrogen for several minutes then held against the dorsal fin of the dolphin for 15 seconds. Ginger has been assigned number 211. Females are given odd numbers; males get even numbers. This allows researchers to quickly recognize the sex of a numbered dolphin they may be watching.

Ginger didn't seem to feel a thing when the brand was pressed against her fin. She was very calm and continued to breathe normally. The number was rinsed repeatedly with clean water. 211 will show up as a white number against her grey skin. After a few months, the brand will be less obvious, but it may be visible throughout her life.

Once the freeze-branding was complete, Ginger was again lifted into the pool. Volunteers and staff moved her out of the stretcher then to the middle of the tank. 3-2-1-go! As Ginger was released, people moved quickly to the walls of the tank, up the ladder, and out. It was exciting to know that in a couple weeks the dolphin would be back out in Sarasota Bay with her friends and family. In the meantime, she swam quietly around the pool.

RELEASE DAY

Ginger was luckier than the patients at Mote who come from further away. On Monday, February 9, she was released into Sarasota Bay, just across the street from the Dolphin and Whale Hospital.

Marine mammals are released as close as possible to the place they stranded, in the hope that they will recognize the habitat and return to friends and family. When the release site is far from Mote, transportation usually takes place at night so there are fewer problems with heat and traffic. Because of this timing, the release of a dolphin is rarely a media event. It's just too dark outside. Ginger, however, was a star! The media were present as she was caught and lowered into the sling one last time. After weighing her, the cables were switched and she was raised over the hospital wall.

As volunteers and staff raced down the stairs and into the truck, Ginger was lowered into place. Randy attached a small radio tag to her dorsal fin. This would be used to track her for as long as the tag stayed in place – usually six to eight weeks.

Even though Ginger was taken less than half a mile, the truck had buckets of water and sponges to keep her wet as she was moved. As with any hospital procedure, one person called "blow" when Ginger breathed, and another person recorded the time of each breath.

There was a crowd at Mote's Chickee Hut, and well-wishers were enjoying coffee and donuts as the truck backed down to the beach. Staff and volunteers lowered the sling and carried Ginger toward the water. The crowd watched expectantly. Did they think she would wave "goodbye" or whistle a gracious "thank you?"

Probably not. They were just hoping that all would go smoothly and that the eight weeks of care would end well. As the sling was carried deeper into the water, the sides were lowered and suddenly Ginger was gone. The cameras were poised; binoculars were set, but there was nothing to see. After several minutes people started to leave. It was over. She was gone.

Out in the bay, a boat was set up with an antenna to track Ginger for the first few hours to make sure she was OK. From the angle of the antenna, it appeared that the boat staff was getting a signal, but from the shore, onlookers saw no sign of Ginger.

Suddenly a dolphin leaped into the air. People still on the beach turned their cameras back on and raised their binoculars. Another leap. Yes, it was Ginger! They could see the 211 on her dorsal fin. She leaped again and yet again.

FOLLOW-UP

Following NMFS regulations, SDRP tracked Ginger via the radio antenna for 60 days after her release.

Shortly after the 60th day, the survey boat spotted Ginger near Roberts Bay. The radio tag was gone. It is made to fall off after the end of the battery life so the dolphin does not "wear" it forever. It had done just what it was supposed to do and Ginger was now free of her tracking device.

From now on, F211 will be just one more Sarasota Bay dolphin – except to the staff and volunteers who nursed her back to health during the cool winter months of 2008 - 2009. To them Ginger will always be special!

ITS A JUNGLE OUT THERE!

Sarasota Bay can be a dangerous place for a dolphin, especially a young animal, recently separated from her mother, who has not had a lot of experience on her own.

While the number of sharks off the Gulf coast of Florida has declined in recent years, probably due to over-fishing, Dr. Wells and his team still find a large number of Sarasota Bay dolphins with evidence of shark bites.

Fewer sharks have probably also meant an increase in the number of stingrays, a favorite shark food. But the rays present another danger to dolphins.

A bottlenose dolphin cruising along the bottom or nosing into the sand to look for food can cause a stingray to launch its barb with a sometimes fatal result. The initial sting rarely causes death, but the barb can work its way into tissues or organs and cause infection or disease.

Then there are the human threats. As we are becoming more and more conscious of the problem of pollution, the water in Sarasota Bay has been improving in quality in recent years. Pollution, however, is still an important factor in the health of marine mammals.

This dolphin's nickname is RipTorn.

A more visible kind of pollution is human garbage – plastic bags and bottles, fishing hooks and line. The plastic can be swallowed and may lodge in the animal's esophagus or stomach, blocking its ability to digest food. Fishing line can tangle around the animal. If it is not cut off or somehow dislodged, the line may infect or sever a fin or fluke.

In a seemingly funny, but really terrrible incident, the SDRP staff

spotted Scrappy, an 8-year-old dolphin, with something that looked like fabric wrapped around him. After several weeks of observation, when the dolphin was still entangled in the fabric,

Dr. Wells got permission from NMFS to catch him. The rescuers found that the material was a large men's Speedo bathing suit, and it was cutting into the base of Scrappy's flippers.

Scrappy, wearing a men's swimsuit

Dr. Wells and his team removed the swimsuit, gave the animal antibiotics to ward off infection, and released him back into the wild. Without their intervention, it is likely that Scrappy would have died from his encounter with a human swimsuit.

Boats present another danger to dolphins. While their hearing is excellent and they are usually fast enough to dodge the speeding vessels, numerous Sarasota Bay dolphins nonetheless show signs of boat strikes.

Loss of habitat is another less visible but no less critical danger to a dolphin population. As new homes and resorts are built along the waterways, there are fewer shallow grass areas for dolphins to fish, mate, and raise their young.

To survive in Sarasota Bay, Ginger will have to find ways to deal with all these threats.

WHAT IS STRANDING?

Under the Marine Mammal Protection Act (1972), stranding is defined as an event in the wild where

A marine mammal is dead and is:

On a beach or shore of the United States; or

In waters under the jurisdiction of the United States (including any navigable waters); or

A marine mammal is alive and is:

On a beach or shore of the United States and unable to return to the water;

On a beach or shore of the United States and, although able to return to the water, is in apparent need of medical attention; or

In the waters under the jurisdiction of the United States (including any navigable waters), but is unable to return to its natural habitat under its own power or without assistance.

SARASOTA DOLPHIN RESEARCH PROGRAM

The Sarasota Dolphin Research Program, started in 1970, is the world's longest running study of a dolphin population. Dr. Randall Wells has been involved in the project from the beginning. As a high school student, Randy was fascinated by the ocean and its creatures. He volunteered at Mote Marine Lab and began working with Blair Irvine, a Mote scientist who was studying dolphins.

Today, Randy, a senior conservation scientist with the Chicago Zoological Society, continues to work at Mote. A cooperative effort between Mote Marine Laboratory and the Chicago Zoological Society enables him to conduct his dolphin research, and to supervise and support the research of other scientists.

Randy and several members of his staff can identify most of the over 150 "resident" dolphins in Sarasota Bay by a brief glance at the dorsal fin. Each fin is unique. Some have shark bites; some have boating scars, and some just have nicks or rake marks from tussling with other dolphins.

On occasion, Randy's team brings in scientists from around the globe to do a health assessment of the Sarasota dolphins. This is a catch-and-release program where animals are taken out of the water for a very brief time. An ultrasound may be used to check the thickness of the animal's blubber. Blood samples can indicate the amounts of pesticides or other contaminants carried by the dolphins, and can provide DNA to help confirm the mother and father of a particular animal.

The boats carry a notebook of photos of dorsal fins of Sarasota Bay dolphins. Randy's research documents a span of five generations of Sarasota Bay dolphins and his methodology has been copied by dolphin projects in Argentina, India, New Zealand, Scotland and other countries around the world.

NOTE: During the 2011 health assessment, Ginger's brother, C1272, was assigned the number 264 and given the nickname "Wasabi."

WHO'S WHO?

Sarasota Dolphin Research Program
Dr. Randall Wells – Manager
Jason – Field Coordinator
Aaron – Research Assistant
Jessica – Graduate Student

Mote Marine Laboratory and Dolphin and Whale Hospital
Dr. Andy Stamper – Consulting Veterinarian
Lynne – Certified Vet Technician; Vet Lab Manager
Connie – Animal Care Technician
David – Animal Care/Nutrition Coordinator
Dr. David Mann – Adjunct Scientist – University of South Florida

PHOTO CREDITS

Ruth Petzold: back cover author photo
Bruce Camardo: p. 21
Chicago Zoological Society: pp. 2-3, 17 (l.), 27, 29, 38 (bottom), 39
Tom Creel: p. 26
Marc M. Ellis / H2O Pictures: p. 16
David Mann (USF) p. 25
Mote Marine Laboratory: pp. 6, 8, 19
Cathy Marine: all others

Thanks to Dr. Randy Wells for advice and editing, to Randy and his staff and the Chicago Zoological Society for photos, and to volunteers and staff at the Dolphin and Whale Hospital, who encouraged me to write this book.   C. M.

GLOSSARY

Cetacean	a whale, dolphin, or porpoise
Dorsal fin	the fin behind the blowhole on the back of the dolphin
Echolocation	also called "sonar;" sound used to locate and identify an object. The dolphin emits clicks that reflect off the object and are heard by the dolphin.
Fluke	the tail fin
Melon	the rounded "forehead" of a dolphin. Important to focusing sounds
Pectoral fins	the flippers or fins on either side of the dolphin. The bones resemble human arms with hands and fingers.
Rostrum	a beak or snout
Signature whistle	whistle that is unique to each dolphin; like a human name
Toxin	poison or a substance that acts like poison and is naturally produced

ACRONYMS

FB or F	Freeze brand. A way of marking a dolphin for identification.
MMPA	Marine Mammal Protection Act
NMFS	National Marine Fisheries Service – a division of NOAA
NOAA	National Oceanic and Atmospheric Administration
SDRP	Sarasota Dolphin Research Program

BIBLIOGRAPHY

Herzing, Dr. Denise L. *The wild dolphin project: long-term research of Atlantic Spotted Dolphins in the Bahamas*. Jupiter, FL: The Wild Dolphin Project, 2002.

Howard, Carol J. *Dolphin chronicles: the two worlds of Echo and Misha*. New York: Bantam Books, 1995.

Pringle, Laurence P. *Dolphin man: exploring the world of dolphins*. New York: Atheneum Books for Young Readers, 1995.

Reynolds, John E., Randall S. Wells, and Samantha D. Eide. *The bottlenose dolphin: biology and conservation*. Gainesville: University of Florida Press, 2000.

Reynolds, John E. III and Randall S. Wells. *Dolphins, whales and manatees of Florida: a guide to sharing their world*. Gainesville: University of Florida Press, 2003.

WEBSITES

www.mote.org Mote Marine Laboratory in Sarasota, Florida. Check on current patients at the hospital.

www.sarasotadolphin.org The Sarasota Dolphin Research Program, including current projects in this 40-year study.

www.wilddolphinproject.org Find out about spotted dolphins in the Bahamas.

www.dolphincommunicationproject.org Learn more about how dolphins communicate.

www.dolphins.org About dolphins at the Dolphin Research Center in the Florida Keys.

www.czs.org The Chicago Zoological Society. Find out about animal research and conservation.

INDEX